PARKINSON'S DISEASE
Dopamine Metabolism, Applied Biochemistry and Nutrition

Lucille Leader and Dr Geoffrey Leader

Foreword by Dr Nicholas Miller

dp

DENOR PRESS

PARKINSON'S DISEASE
Dopamine Metabolism Applied Biochemistry and Nutrition
ISBN 978 0 9526056 6 9

British Library Cataloguing in Publication Data.
A catalogue record for this book is available from the British Library.

Published by Denor Press Ltd. UK:
Website: www.denorpress.com

Cover Design by Commercial Campaigns from an idea by Dr Geoffrey Leader.

Layout and Design by Commercial Campaigns, UK

Lucille Leader Dip ION MBANT
email: denorgroup@gmail.com

Lecturer in Nutritional Therapy at Westminster University,
School of Integrated Health, London, UK

Nutrition Director, The London Pain Relief and Nutritional Support Clinic,
The Highgate Hospital, London, UK

Council Member, Food and Health Forum, Royal Society of Medicine, London, UK

Member of the British Society for Ecological Medicine, London, UK

Lectured in Europe for The European Parkinson's Disease Association (EPDA) in Specialised Nutritional Management in Parkinson's Disease, in the USA for the Parkinson's Resource Organisation (Meeting of the Minds), in Vienna at the first Congress for Sexuality and Nutrition in Parkinson's Disease, in South Africa for the Johannesburg Parkinson's Disease Society and at UK Parkinson's Disease Societies. She has received a "Quality of Life in Parkinson's Award" in the USA and the CAM "Outstanding Practice Award" for the UK. Amongst other publications as author, Lucille Leader is co-author of six successful books on Parkinson's Disease.

Geoffrey Leader MB ChB FRCA
email: denorgroup@gmail.com

Medical Director, The London Pain Relief and Nutritional Support Clinic,
The Highgate Hospital, London, UK

Consultant Anaesthetist, The Highgate Hospital, London, UK
The Wellington Hospital, London, UK

Formerly Chairman of the Anaesthetic Department,
Senior Anaesthetic Consultant, Senior Lecturer, Pain Clinic Consultant,
Director of Intensive Care Unit, Newham General Hospital
(The London Hospital Medical College, London University, UK)

Lectured in Europe for The European Parkinson's Disease Association (EPDA) in Specialised Nutritional Management in Parkinson's Disease, in the USA for the Parkinson's Resource Organisation (Meeting of the Minds), in Vienna at the first Congress for Sexuality and Nutrition in Parkinson's Disease, in South Africa for the Johannesburg Parkinson's Disease Society and at UK Parkinson's Disease Societies. He has received a "Quality of Life in Parkinson's Award" in the USA. Geoffrey Leader is the co-author of six successful books on Parkinson's Disease. He has contributed to peer-reviewed anaesthetic journals and is especially interested in the nutritional support of patients with acute and chronic illness.

i

Dr Nicholas Miller MA MSc PhD CSci MCB FRCPath
email: n.j.miller@biolab.co.uk

Nicholas Miller holds science degrees from Trinity College, Dublin and completed his PhD at the University of London. He has worked in the field of Clinical Chemistry for 30 years, initially as an endocrine biochemist, subsequently developing an interest in nutritional medicine. He was formerly Senior Lecturer in Biochemistry in Guy's Hospital Medical School, where he worked with Professor Tony Diplock (the leading authority on vitamin E) on a number of projects, including the Whitehall Survey. Most of his publications in the medical literature (approximately 70 papers, including those for The Royal Society of Chemistry, Journal of Diabetes Complications, Free Radical Research and Free Radical Biology and Medicine) concern the role of free radicals and antioxidants in the diet and in disease.

He has been Laboratory Director of Biolab Medical Unit Laboratory Director of Biolab Medical Unit in London UK, which is a nutritional and environmental laboratory specialising in elemental analysis, vitamins and nutritional enzymology.

Foreword

During the normal ageing process, oxidative damage to the brain increases as levels of neurotransmitters fall. The brain, which in the human consumes about 20% of the inspired oxygen in a resting individual, contains mainly polyunsaturated fatty acids, and as defence mechanisms diminish in their activity, neural tissue is left open to a process of on-going oxidative damage. Many of these defence mechanisms depend on the intake of certain micronutrients.

Age-related brain deterioration, such as that of Parkinson's disease, is generally thought to be related, amongst other reasons, to the damage inflicted by oxygen free radicals and their intermediates. In view of this, the possibility of deferring age-related degenerative changes in the brain has become a major interest.

Lucille and Geoffrey Leader, who are well known and much admired for their work over many years with sufferers from Parkinson's disease, have uniquely applied the principles of individualised biochemically-based nutritional science in their clinical practice. They present in this book a modern, up-to-date view of the biochemical background of Parkinson's disease.

It includes support of biochemical pathways pertinent to dopamine metabolism, mitochondrial stimulation, oxidative stress, status of the gastro-intestinal tract, glucose regulation, drug nutrient-interactions, detoxification, methylation and adrenal aspects. All of these factors are of concern in the presentation of Parkinson's Disease. These protocols may provide strategies for the amelioration of symptoms and affect the progression of the disease.

This book will be of great interest to clinicians, nutritionists and students and also to individuals experiencing the symptoms of Parkinsonism. It is to be hoped that it will also encourage debate and discussion as to the role of nutritional interventions, along with dopaminergic and L-dopa medication, in the treatment and control of this disease.

Nicholas J. Miller MA MSc PhD CSci MCB FRCPath
Laboratory Director
Biolab Medical Unit
London, UK

Dedication

To the Memories of
Dr Erich Segal and Tom Isaacs
Parkinson's Disease Patients
Biochemical Laboratories
Dr Serena Leader, Joe Leader, Felicia Beder,
Michael Beder, Justin Beder, Zack Orton
Morris Orden

Introduction

The science of Biochemistry shows how lifeless molecules take part in the living processes of Metabolism to generate and consume energy. Activation of the enzymes that control these transformations requires nutritional factors, usually minerals or vitamins, which are supplied from the diet.

Metabolic compromises are manifested in Parkinson's Disease. These include inadequacies in the production of the neurotransmitter dopamine (metabolised from dietary protein) and protection from inflammatory responses (prostaglandins are metabolised from dietary essential fatty acids).

Oxidative stress (oxygen- and nitrogen-centred free radicals) is a problem (antioxidant enzymes have specific nutrients as coenzymes). Mitochondrial function (dependent upon glucose and specific nutrients) is also compromised.

Catecholamine response as a result of stress can result in inhibition of the rate - limiting enzyme tyrosine hydroxylase with its nutritional coenzymes.

Therefore, to understand how to support the metabolic and biochemical aspects implicated in endogenous dopamine production, as well as those other cellular needs of the Parkinson's patient, we feel that this analytical information should be introduced, in succinct style, to neurologists and associated health professionals. It can also be of vital interest to those with the disease.

We sincerely hope that it will stimulate the broader use of contemporary biochemical tests to plan individualised adjuvant care for the Parkinson sufferer. This could well enhance the therapeutic spectrum.

Lucille Leader and Geoffrey Leader
Parkinson's Disease Management Clinic
Highgate Hospital
London, UK

Disclaimer

The recommendations in this book are not intended to replace general medical advice or the advice of a neurologist, nutritionist, dietician, clinical biochemist, pathologist, pharmacist, psychologist or any other healthcare professional.

Medical knowledge is constantly expanding. As new experimental and clinical experiences are gained, management needs to be constantly re-evaluated and updated.

In the best and safe interests of patients, the biochemical individuality of patients makes it imperative for their supervision by healthcare professionals at all times.

As such, the publisher and authors cannot accept liability for any problems arising directly or indirectly from the application of the principles presented in this text.

Contents

Parkinson's Disease
Dopamine Metabolism, Applied Biochemistry and Nutrition

The Parkinson's disease sufferer, deficient in the neurotransmitter dopamine, surely deserves to function as well as possible. Therefore, concomitant with dopaminergic pharmaceutical supplementation and comprehensive multidisciplinary care, dynamic support of the individual's biochemical and metabolic status should be considered as adjuvant therapy in order to optimise any potential cellular function during the degenerative journey.

Cardinal biochemical and functional pathways for consideration:

- Dopamine Metabolism and Biochemistry

- Dopamine Metabolism and Adrenal Biochemistry

- Parkinson's Disease and Mitochondrial Function

- Comprehensive Tyrosine Metabolism

- Oxidative Stress

- Detoxification

- Inflammation and Fatty Acid Metabolism

- Environmental Implications

- Intestinal Function and Regulation

- Drug and Nutrient Interactions

- Biochemical Tests

- Nutritional Supplements

Notes:

1. *Dopamine Metabolism*

DIETARY PROTEIN (EGG, FISH, POULTRY, SOY)	- *Hydrochloric acid,* *Carbonic anhydrase (zinc), Proteases*
PHENYLALANINE	- *Phenylalanine hydroxylase*
TYROSINE	- *Tyrosine hydroxylase* *(iron, tetrahydrobiopterin)*
L-DOPA	- *Dopa decarboxylase (vit B6),*
	- *Catechol-o-methyltransferase*
DOPAMINE	- *Dopamine β-hydroxylase (copper, vit C)*
NORADRENALINE	- *Phenylethanolamine N-methyltransferase* *(S-adenosylmethionine)*
ADRENALINE	- *(vit C)*
CORTISOL & DHEA	

Fig 1: *Dopamine Metabolism: Nutritional Biochemistry with Enzymes and Co-enzymes*

The amino acid tyrosine, metabolized from concentrated dietary protein containing the essential amino acids, metabolises to levodopa and on to dopamine.[1] The cascade to the formation of tyrosine is dependent on the following criteria:

➤ Initial digestion by hydrochloride acid, pepsin and carbonic anhydrase (zinc[2] dependent) in the stomach and further by proteases in the proximal small bowel to form tyrosine.

These processes are followed by further enzymatic actions in the periphery:

➤ Tyrosine hydroxylase, whose function is activated by specific coenzymes biopterin[3] (a folate derivative) and iron,[4] metabolizes tyrosine to levodopa.

➤ Catechol-o-methyltransferase and dopa decarboxylase, the latter activated by the specific coenzyme vitamin B6, metabolizes levodopa to dopamine.

5

Notes:

2. Parkinson's Disease and Adrenal Biochemistry

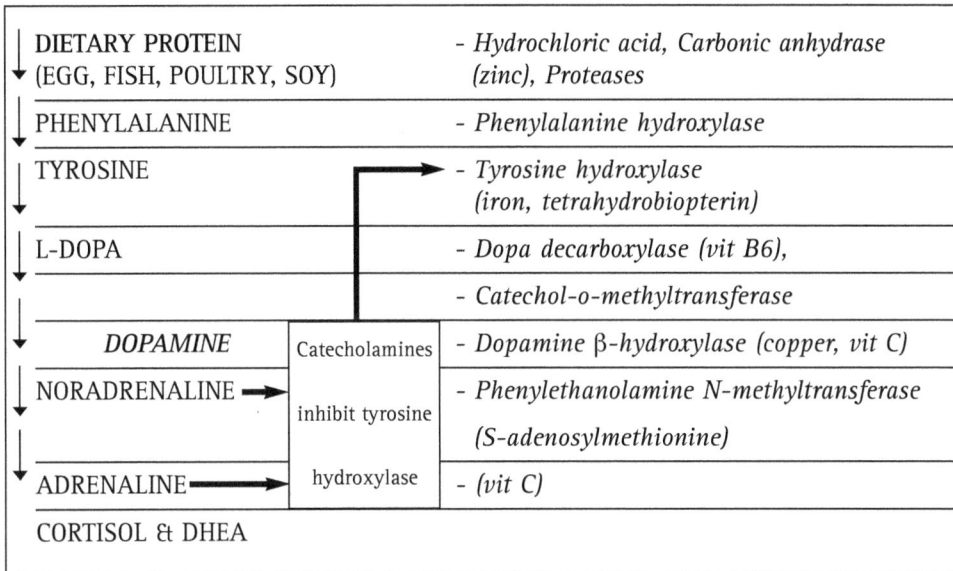

↓ DIETARY PROTEIN ↓ (EGG, FISH, POULTRY, SOY)	- *Hydrochloric acid, Carbonic anhydrase (zinc), Proteases*
↓ PHENYLALANINE ↓	- *Phenylalanine hydroxylase*
↓ TYROSINE ↓	- *Tyrosine hydroxylase (iron, tetrahydrobiopterin)*
↓ L-DOPA ↓	- *Dopa decarboxylase (vit B6),*
↓	- *Catechol-o-methyltransferase*
↓ *DOPAMINE*	- *Dopamine β-hydroxylase (copper, vit C)*
↓ NORADRENALINE →	- *Phenylethanolamine N-methyltransferase*
↓	*(S-adenosylmethionine)*
↓ ADRENALINE →	- *(vit C)*
CORTISOL & DHEA	

Catecholamines inhibit tyrosine hydroxylase

Fig 1a: *Dopamine Metabolism: Nutritional Biochemistry with Enzymes and* Co-enzymes

Physical symptoms of Parkinson's disease are often exacerbated by stressors. Adrenaline, released in response to stress, is metabolised from dopamine. Dopamine metabolises to noradrenalin[5] (enzyme dopamine B-hydroxlase with coenzymes vitamin C and copper) and further on to adrenaline[6] (phenylethanolamine N-methyltransferase with coenzyme s-adenosylmethionine (SAM-e)). However, catecholamine feedback adversely affects tyrosine hydroxylase.[7] There is feedback inhibition by the catecholamines, which compete with the enzyme for the pteridine cofactor. Depression[8] and anxiety, with the adrenal consequences, may also be experienced by patients with Parkinson's disease.

The nutrients directly involved in the metabolism from dietary protein, on to tyrosine, through to dopamine, noradrenaline and adrenaline include vitamin C, vitamin B6, biopterin, iron, copper and methionine[9] (see figure 1). The adrenal glands are nutrient dependant, including vitamin B6,[10] vitamin B5,[11] and vitamin C.[12] Inordinate DHEA and cortisol output in response to chronic unremitting stress may be causative of other pathologies, disrupts blood sugar regulation and when cortisol levels remain low, may also lead to adrenal exhaustion.[13]

3. Parkinson's Disease and Mitochondrial Function

The citric acid cycle (Krebs Cycle) which produces cellular energy is compromised in Parkinson's disease.[14]
L-dopa administration further down-regulates the Krebs Cycle.[15,16] Adenosine triphosphate (ATP) produced by the cycle is dependent on glucose and specific vitamins and minerals for its metabolism.[17]

Glucose is metabolised from carbohydrates (fruit and vegetables) and if this is not available protein and fat stores are metabolised for use in the cycle by the process of gluconeogenesis.

The enzyme acetyl-CoA carboxylase is biotin dependant.[18] Carnitine[19] is also involved.

The Krebs Cycle[20] depends on specific vitamins and minerals which include amongst others vitamins B1, B2, B3, B5, magnesium, manganese, vitamin C, copper, iron and co-enzyme Q-10. Oxygen is necessary to the production of ATP.

FATS **CARBOHYDRATES** **PROTEINS**

STAGE I — DIGESTION & ASSIMILATION

Fatty Acids, Glycerol Cholesterol   Glucose & Other Sugars   Amino Acids

Keto Acids

Pyruvate ⇌ *Lactate*

Adipate
Suberate
Ethylmalonate

| Carnitine | B_1, B_2, B_3, B_5, Lipoate | B_1, B_2, B_3, B_5, Lipoate |

Acetyl CoA

→ β-**Hydroxybutyrate**

STAGE II — INTERMEDIARY METABOLISM

Citric Acid Cycle

Asp → **Oxaloacetate**   **Citrate**

B_3 **cis-Aconitate**

Malate | Cysteine, Fe++ |

| Glu His Arg Pro Gln |

Tyr Phe → **Fumarate**   **Isocitrate**

NADH | B_3, Mg, Mn |

B_2 *NADH*   **α-ketoglutarate**

NADH | B_1, B_2, B_3, B_5, Lipoate |

Succinate | Mg | **Succinyl-CoA** ←

| Leu Ile Val Met |

NADH

FADH$_2$

STAGE III — ELECTRON TRANSPORT AND OXIDATIVE PHOSPHORYLATION

NADH Dehydrogenase

— ADP + P_i

| Coenzyme Q_{10} |

Biosynthesis

Hydroxymethylglutarate (HMG)   Cytochromes

O_2 H_2O

ATP

energy
(Muscle, nerve function, maintenance, repair)

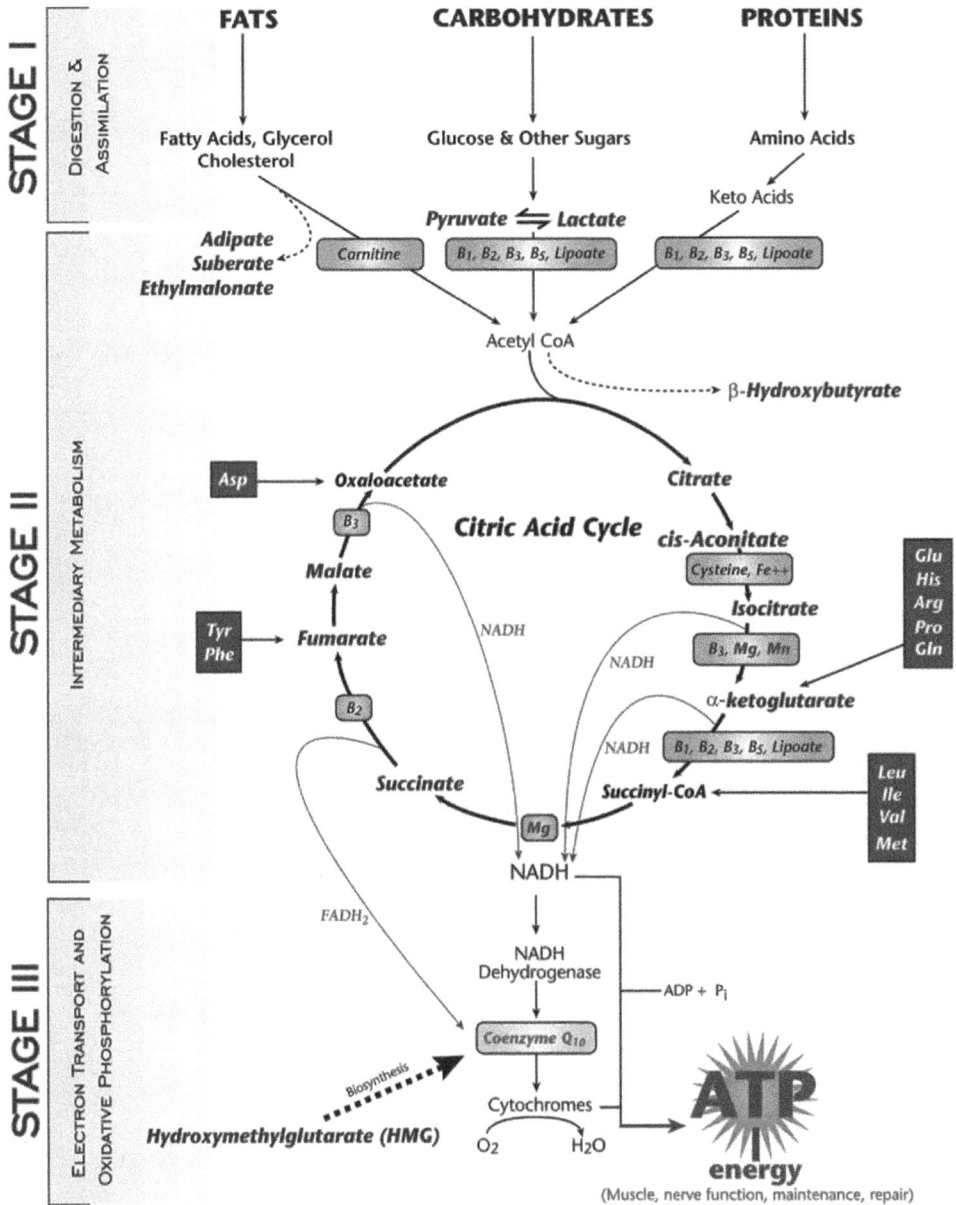

Fig 2: *Nutritional Cofactors in the Citric Acid Cycle*[20]

With permission from Bralley JA & Lord RS. Organic Acids in Urine. Laboratory Evaluations for Integrative and Functional Medicine 2nd Edition, Metametrix Institute 2008;6:325

Notes:

4. *Tyrosine Metabolism*[21,22]

Tyrosine, the amino acid metabolised from dietary protein, is the precursor not only of dopamine but also of thyroxine (T4), T3 and melanins. Metabolism to thyroxine (T4) and T3 includes iodine[23] as well as iron and selenium.[24] Metabolism from tyrosine to melanins is dependent on copper.[25]

With tyrosine being the rate limiting amino acid for dopamine metabolism, one could postulate that in some individuals there may be metabolic disturbance between tyrosine and thyroxine as well as to melanins.

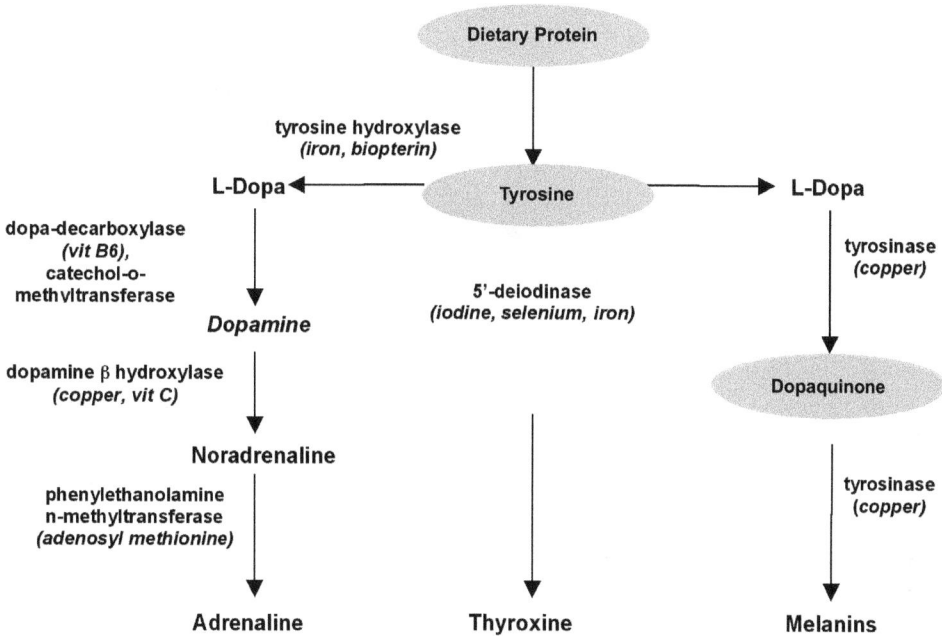

Fig 3: *Tyrosine Metabolism to Adrenaline, Thyroxine and Melanins*

5. Oxidative Stress

In Parkinson's disease there is inordinate oxidative stress[26] and free radical production. This occurs when reactive oxygen products produced are beyond the coping mechanisms of nature.

Impaired oxidation of pyruvate has been demonstrated in human embryonic fibroblasts after exposure to L-dopa.[27]

Physiologically, free radical activity is quenched by antioxidants.[28] These include (amongst others) glutathione, vitamin C, vitamin E, lipoic acid and selenium. Antioxidant enzymes are glutathione peroxidase (selenium dependent) and superoxide dismutase (dependent on zinc, copper, manganese) and catalase (zinc dependent).

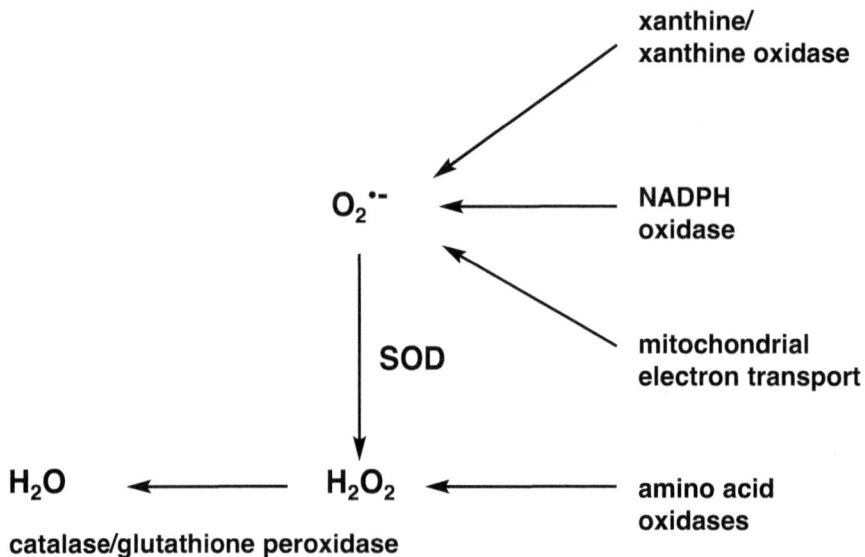

Figure 4: Synthesis of Free Radicals in the Body.
With permission from Dr Nicholas Miller (diagram & description), Biolab Medical Unit, London UK

The normal output of superoxide (the single electron reduction product of oxygen) has been estimated at 1 kilogram per year and is derived from pathways including mitochondrial electron transport. Free radicals are chemical species with an unpaired electron in their outer orbital and are hence highly reactive.

$O_2^{\bullet-}$ ___ $NO^{\bullet}$ → $OONO^-$ —————→ lipid damage
protein modification
antioxidant consumption (loss)

iron
copper → $^{\bullet}OH$ → lipid peroxidation

→ cholesterol oxidation

H_2O_2

haem
protein → DNA damage

ferryl
radical → antioxidant consumption (loss)

HOCl —————————————→ protein modification
cholesterol chlorination

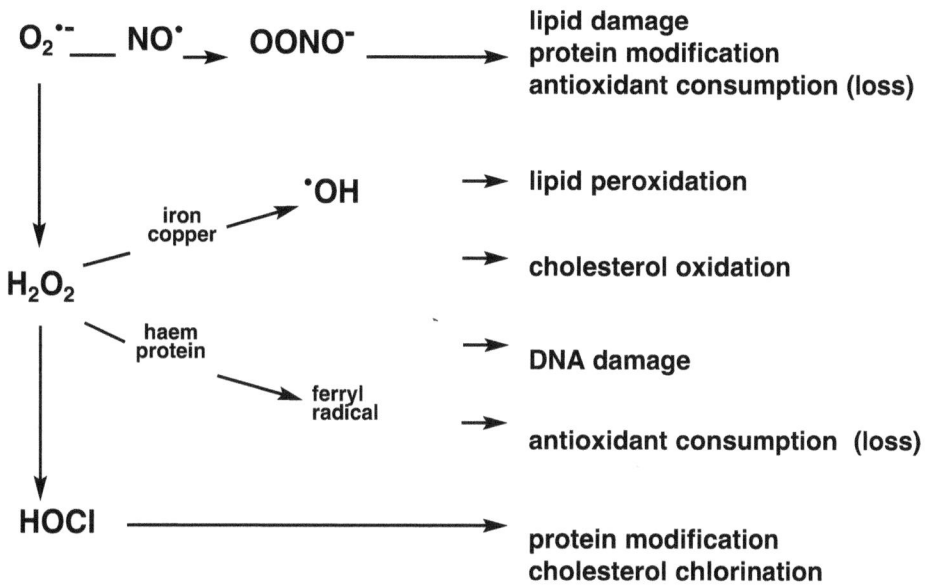

Figure 5 - Damage to Tissues by Free Radicals.
With permission from Dr Nicholas Miller (diagram & description), Biolab Medical Unit, London UK

In oxidative stress the protection afforded by endogenous antioxidant enzymes is overwhelmed. The secondary radicals produced may be very damaging to every tissue in the body. Peroxynitrite, the product of superoxide and nitric oxide (NO), is a highly-damaging radical which also has the ability to cross lipid membranes (and hence can pass into the brain).

Peroxynitrite may also be generated in sites where there are vascular changes (release of NO) and also tissue damage (leakage of superoxide.) Loss of magnesium and membrane fatty acids are among the easily-observable consequences of oxidative stress.

Notes:

6. Detoxification

In Parkinson's disease, the liver's ability to detoxify efficiently may be impaired.[29]

There are two phases of liver detoxification[30]:

Phase 1:   Many drugs and nutrition itself require Phase1 detoxification with efficient P450 enzymes. Nutrients pertinent to Phase 1 include vitamins B2, B3, B6, B12, folic acid, glutathione, branched-chain amino acids, flavonoids and phospholipids.

Phase 2:   Nutrients pertinent to Phase 2 liver detoxification are glycine, taurine, glutamine, n-acetylcysteine, cysteine, methionine.

Glutathione production relies on dietary protein, metabolised from the amino acid methionine. Homocysteine levels depend on the adequacy of the methylation process[31] which depends on vitamins B12, B6 and folate. Raised homocysteine[32] levels may be found in Parkinson's disease as well as in cardiovascular disease.

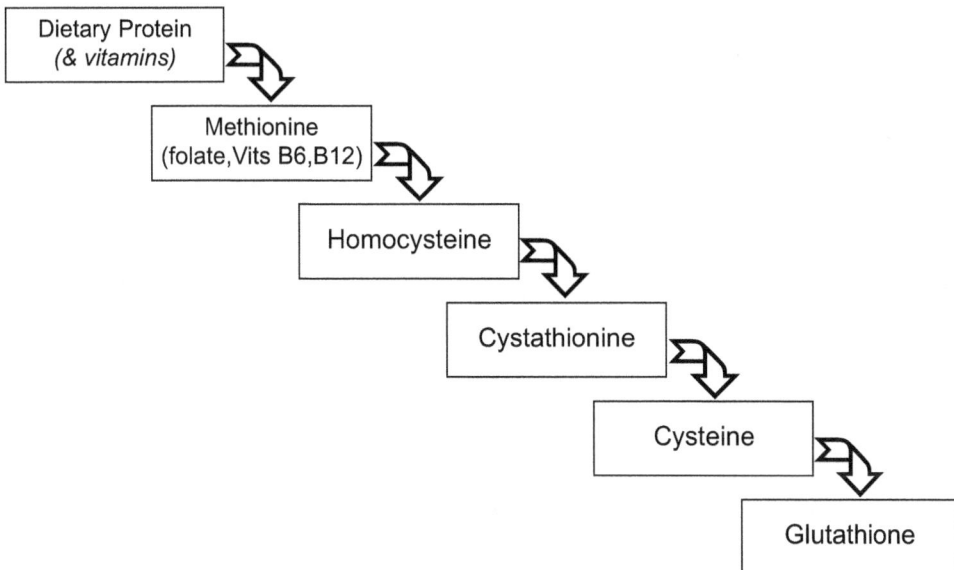

Fig 6: *Glutathione Metabolism with Methylation*

15

7. Inflammation and Fatty Acid Metabolism

Parkinson's disease is associated with increased inflammation at cellular level.[33] Physiological modification of inflammation depends upon the anti-inflammatory prostaglandins[34] Prostaglandin Series 1 and Prostaglandin Series 3. These are both metabolised from essential fatty acids in the diet by the enzymes[35] delta-6-desaturase with coenzymes vitamin B6, zinc and magnesium and delta-5-desaturase with vitamins B3, C and zinc.

Omega 6 (Prostaglandin Series 1) is metabolised from nuts and seeds and Omega 3 (Prostaglandin Series 3) from oily fish. Delta-4-desaturase[36] is also implicated in these pathways.

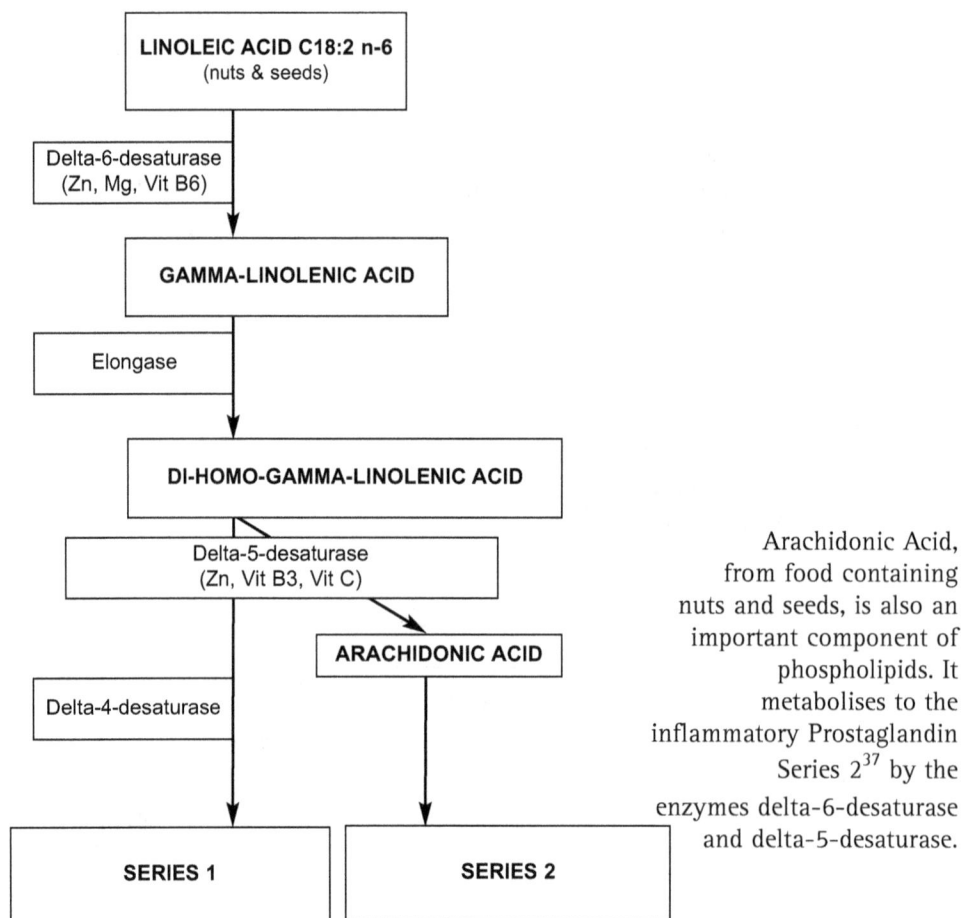

LINOLEIC ACID C18:2 n-6
(nuts & seeds)

Delta-6-desaturase
(Zn, Mg, Vit B6)

GAMMA-LINOLENIC ACID

Elongase

DI-HOMO-GAMMA-LINOLENIC ACID

Delta-5-desaturase
(Zn, Vit B3, Vit C)

ARACHIDONIC ACID

Delta-4-desaturase

SERIES 1

SERIES 2

Arachidonic Acid, from food containing nuts and seeds, is also an important component of phospholipids. It metabolises to the inflammatory Prostaglandin Series 2[37] by the enzymes delta-6-desaturase and delta-5-desaturase.

Fig 7: *Essential Fatty Acid Metabolism from Linoleic Acid*

The Essential Fatty Acids, critical to life, are metabolised exclusively from dietary sources. They produce the ecosanoids which in turn give rise to prostaglandins as well as prostacyclins, thromboxanes, leukotrienes and lipoxins.

They are found in the structural lipids of cells, in phospholipids as well as playing a role in the structural integrity of the mitochondrial membrane.

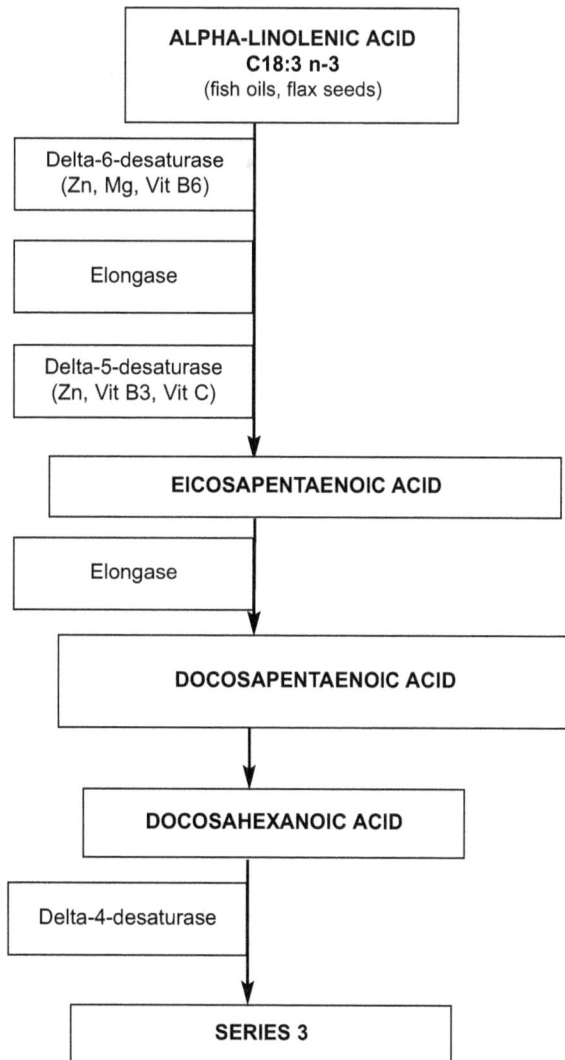

```
                          ┌─────────────────────────────┐
                          │     ALPHA-LINOLENIC ACID      │
                          │          C18:3 n-3            │
                          │     (fish oils, flax seeds)   │
                          └─────────────────────────────┘
              ┌────────────────────────────┐
              │   Delta-6-desaturase        │
              │   (Zn, Mg, Vit B6)          │
              └────────────────────────────┘

              ┌────────────────────────────┐
              │        Elongase             │
              └────────────────────────────┘

              ┌────────────────────────────┐
              │   Delta-5-desaturase        │
              │   (Zn, Vit B3, Vit C)       │
              └────────────────────────────┘
                          ┌─────────────────────────────┐
                          │    EICOSAPENTAENOIC ACID      │
                          └─────────────────────────────┘
              ┌────────────────────────────┐
              │        Elongase             │
              └────────────────────────────┘
                          ┌─────────────────────────────┐
                          │   DOCOSAPENTAENOIC ACID       │
                          └─────────────────────────────┘
                          ┌─────────────────────────────┐
                          │   DOCOSAHEXANOIC ACID         │
                          └─────────────────────────────┘
              ┌────────────────────────────┐
              │   Delta-4-desaturase        │
              └────────────────────────────┘
                          ┌─────────────────────────────┐
                          │         SERIES 3              │
                          └─────────────────────────────┘
```

Fig 8: *Essential Fatty Acid Metabolism from Alpha-Linolenic Acid*

Notes:

8. *Intestinal Function and Regulation*

As dopamine is metabolised from dietary protein, assessment of digestive enzyme potential (hydrochloric acid, proteases, lipases) is indicated.

Assessment of intestinal flora environment may demonstrate any dysbiosis which could potentially impact on gut immunity.

The degree of intestinal permeability is also an important assessment as in susceptible patients, partly digested protein molecules may cause disruption of the paracellular tight junctions[38] in the gut mucosa. Increased permeability could allow for the absorption of potential pathogens into the circulation. A high protein content food, the peanut, contains an anti-trypsin factor and is therefore difficult to digest. The ensuing incompletely digested protein molecules may increase gut permeability.

There are dietary influences on cellular response, such as food which contains lectins and gluten.

Because of their resistance to digestive, proteolytic breakdown, the luminal concentration of lectins can be quite high. Consequently their transport through the gut wall can exceed that of other dietary antigens by several orders of magnitude.[39]

Lectins (wheat germ agglutinin (WGA) and phytohemagglutinin (PHA) from kidney beans) increase gut permeability by altering tight junction characteristics.[40]

This increase may allow passage of normal gut bacteria, pathogens and antigenic partially-degraded dietary proteins into systemic circulation, causing persistent peripheral T-cell stimulation.

It has also been shown that WGA can bind (in vitro) the following tissues and organs: alimentary tract (mouth, stomach, intestines), pancreas, musculo-skeletal system, kidney, skin, nervous and myelin tissues, reproductive organs, platelets and plasma proteins.[41]

Gluten intolerance[42] and coeliac disease are a consideration in Parkinson's Disease.

Casein and lactose intolerance may be pertinent.

Red meat, high in protein, requires a relatively lengthy time for digestion and transit through the intestinal tract. This may be a consideration in diet planning for patients with chronic constipation.

Patients with Parkinson's disease often experience chronic constipation. Daily, gentle regulation of bowel function is important to ensure optimal absorption of drugs and nutrients, as well as reducing stress. Chronic use of harsh laxatives could lead to melanosis and other disturbances in the intestinal tract.

The Enteric Nervous System (ENS) has been called the Second Brain as there are numerous neurones in the intestinal tract.[43]

Consideration of The Gut-Brain Axis is a vital aspect in the therapeutic approach.

9. *Environmental Implications*

Radiation,[44] organophosphates[45,46] and mercury[47] have been associated with neuronal damage.

Amongst food additives, excitotoxins such as aspartame[48] and monosodium glutamate[49] have been shown to cause neuronal problems at cellular level.

10. *Drug and Nutrient Interaction*

L-dopa competes for absorption with the large neutral amino acids (from dietary protein) at the receptor sites in the intestinal tract, as well as at the blood brain barrier.[50]

For optimal drug response, drug management should include manipulation of the time of taking L-dopa medication in relation to the intake of food containing the large neutral amino acids.[51]

Transit time for stomach emptying as well as the digestive enzyme function of each individual patient should be assessed in order to plan an optimum dietary and drug program which avoids negative drug-nutrient interactions.

Patients who are taking drugs containing monoamine oxidase (MAO) inhibitors at high doses run the risk of hypertension if eating tyramine-rich foods in conjunction with them.[52]

If L-dopa pharmaceuticals do not contain a decarboxylase inhibitor (DDI), therapeutic doses of vitamin B6 are contraindicated for patients. Strict monitoring of low doses away from L-dopa monotherapy is necessary. However, the addition of a dopa decarboxylase inhibitor (carbidopa or benserazide) to L-dopa pharmaceuticals such as "Sinemet", "Madopar" and "Stalevo", prevents peripheral levodopa transformation to dopamine facilitated by pyridoxine before L-dopa crosses the blood brain barrier (BBB). This ensures efficacy of the drug.

It is therefore possible, if patients are taking L-dopa concomitantly with a DDI, to be prescribed vitamin B6[53,54] if indicated. Indeed, it is advisable to evaluate the status of this important nutrient. It functions as the coenzyme with dopa decarboxylase in the metabolic step from L-dopa to dopamine[55,56] as well as playing a role in other anabolic and metabolic functions.

Iron may interact with L-dopa[57] ("Madopar" and "Sinemet") as well as with Stalevo[58] and Entacapone "Comtess"[59]. Antacids reduce the absorption of L-dopa[60].

Alcohol may negatively interact with Pramipexole "Mirapexin"[61].

The manufacturers of Ropinerole "Requip" also state that patients should be cautioned against taking alcohol with centrally active medication[62].

Note: Whilst this section recommends checking for drug-nutrient interactions, it is prudent to assess the possibility of drug-drug interactions as well as the possibility of patient intolerance to individual drug ingredients, including chemical, nutritional and other additives. There are also contraindications to the administration of dopaminergic drugs during pregnancy and breast feeding.

11. *Conclusion*

Dopaminergic drugs are effective in supporting movement disturbance associated with dopamine deficiency - mono-therapy (dopaminergic pharmaceuticals) is still the norm at most clinics. It is evident, however, that mono-therapy of any kind is limiting in the specialized general management of Parkinson's disease.

It can be postulated that it is essential to assess the biochemical prerequisites for endogenous dopamine metabolism, mitochondrial function as well as antioxidant status and oxidative stress, inflammatory markers, intestinal status and the adrenal stress index. This is to provide comprehensive support of metabolic and functional health in Parkinson's disease, as well as enhancing the therapeutic spectrum. In addition to routine medical tests, contemporary biochemical assays are available to demonstrate the biochemical, nutritional and metabolic status of cells, for consideration of up or down regulation by the clinician.

With tests supplying each patient's biochemical individuality, functional health may be optimised within the constraints of the disease.

It is the authors, empirical, clinical experience that patients who have biochemically-based nutritional therapy, concomitantly with dopaminergic drug therapy, experience a better sense of wellbeing than those who are on mono-therapy. Those patients who are not yet on drug therapy also benefit from specialized nutritional management based on their individual biochemistry.

A sample of 100 patients with Parkinson's disease at the authors, clinic demonstrated deficiencies of one kind or another pertinent to dopamine *metabolism and mitochondrial function.*

It is hoped that this text will stimulate further in-depth studies into the biochemical individuality of patients with Parkinson's disease, thus optimizing clinical practice.

This may demonstrate that biochemically-based nutritional support, combined with drug therapy, should become a necessary part of Parkinson's disease management alongside paramedical care.

Notes:

13. *Biochemical Tests*

Assessing patients on a cellular level enables healthcare professionals to plan patient management based on biochemical individuality. Some of the following tests may be useful tools in the assessment of metabolic, nutritional and general health status.

- **Adrenal Stress Index**
  This test is a measure of cortisol and DHEA levels and is used as a biochemical marker of stress. Alteration of the levels of these hormones may have beneficial therapeutic effects.

- **Amino Acid Profile**
  L-dopa is metabolized from dietary protein. It is useful to assess the general amino acid status of Parkinson's Disease patients. Amino acid supplements may be indicated.

- **Antioxidant Profile**
  Oxidative stress may play a role in Parkinson's Disease. Antioxidant therapy may be indicated.

- **Biotin**
  This is a consideration for mitochondrial energy production.

- **Brain Neurotransmitters**
  Measurement of levels of neurotransmitters.

- **Co-enzyme Q10**
  This nutrient plays a vital role in cell energy production. It is also important in cardiovascular health.

- **Digestive Enzymes**
  Patients with Parkinson's disease may demonstrate deficiencies in the production of hydrochloric acid and/or pancreatic exocrine function. As dopamine is metabolized from protein it is essential that dietary protein be adequately digested to initiate this metabolic pathway. The digestion and absorption of carbohydrates and fats are essential to metabolism. Incompletely digested food molecules, particularly large protein molecules, may contribute to increased permeability of the gut mucosa.

- DNA Adducts

- Essential Fatty Acids
  Essential Fatty Acids are essential to the nervous system, production of inflammatory and anti-inflammatory prostaglandins, skin integrity and immune function.

- Folate
  Folate is involved in the control of homocysteine and catecholemine synthesis and is linked with tetra-hydrobiopterin levels.

- Food Allergy (IgE) and Food Intolerances
  It has been observed clinically that certain foods and chemicals may aggravate symptoms in patients with Parkinson's Disease.

- Glutathione
  This is a powerful antioxidant and implicated in liver detoxification

- Glutathione Peroxidase
  This is an antioxidant enzyme with selenium as coenzyme.

- Gut Permeability
  The integrity of the intestinal mucosa is essential for optimal nutritional absorption and protection against the entry of potential pathogens and allergens into the circulation. Permeability could be increased by disease, drugs, surgery, radiation and incompletely digested large protein molecules.

- Haematology, Biochemistry and Ferritin
  General screens investigating immune function, response to infection, electrolytes, kidney function, liver function, haemoglobin, iron and lipid status.

- Hormones Male / Female
  Patients may present with sexual, fertility, menstrual and menopausal problems.

- Liver Detox Test (Phase 1, 2)
  Phase 1 liver detoxification in patients with Parkinson's disease has been shown to be sub-optimal. It is important to assess both phase 1 and phase 2 detoxification pathways in order to optimize nutritional management.

- **Metabolic Analysis Profile (Organic Acids)**
  Analysis of metabolic factors including the citric acid cycle (cell energy), neurotransmitter metabolites, markers of intestinal malabsorption and dysbiosis and co-factor dependent metabolites from amino acid catabolism.

- **Minerals**
  Minerals are essential as co-enzymes. Some examples: zinc plays a role in protein metabolism, magnesium influences relaxation of muscles and chromium is essential to glucose tolerance factor (GTF).

- **Parasites**
  Parasites may be a problem for patients who present with chronic diarrhoea, fatigue, anal irritation and other symptoms.

- **Superoxide Dismutase Detail Studies**
  This is an antioxidant enzyme with zinc and copper as coenzymes.

- **Thyroid Function (TSH, T4, T3, Auto-antibodies)**
  Tyrosine metabolizes three ways – to form L-dopa, thyroxine and melanins.

- **Toxic Metals / Pesticides Screens**
  Patients may have been overly exposed to organophosphates and toxic metals, which affect neurones.

- **Vitamins**

  Vitamin A, Carotenes

  B Vitamins 1,2,6
  Vitamins B1 and B2 are essential to the Citric Acid Cycle and the production of cellular energy (adenosine triphosphate, ATP). Vitamin B6 is the essential co-enzyme in the metabolic step from L-dopa to dopamine.

  Vitamin B3
  Forms NADH for cell energy production and is a biopterin coenzyme.

  Vitamin B12
  This is associated with nervous/muscular function and appears to be required for tetrahydrobiopterin synthesis.

  Vitamin E
  This is an antioxidant.

Notes:

14. *Nutritional Supplementation*

Following is a list of commonly used nutritional supplements, with brief comments. Their use should routinely be based on patients' biochemical individuality, always administered under professional supervision. There may be serious contraindications to taking some in individual cases.

However, nutritional deficiencies also occur in people who do not have Parkinson's Disease and their impoverished cellular environment will equally require support.

Oral supplements which are especially designed to be taken sublingually are better absorbed than capsules and tablets.

Tube feeding may be indicated if dysphagia is a problem.

In cases of malabsorption patients may need intravenous nutritional support.

- **N-acetyl-cysteine (Glutathione Precursor)**
  This stimulates the body's own production of glutathione,[63,64] an antioxidant also involved in the liver detoxification process.

  However, in large doses it may act as an excitotoxin. Some practitioners prefer to give glutathione intravenously.

- **Acetyl-L-carnitine**[65]
  Acetyl-L-carnitine is an important antioxidant.

- **Amino Acids / Poly-peptides**
  These forms of pre-digested protein are for anabolic function. Pre-digested amino acids and poly-peptides are more quickly absorbed than protein food, which makes them eminently suitable for those who need L-dopa medication at shorter intervals.

  Protein supplementation may be necessary if patients are on a reduced protein diet.

- **Anti-oxidants**[66,67,68,69,70]
  These are implicated in control of free-radicals and include alpha lipoic acid, vitamins C and E as well as selenium, zinc, copper and gluthathione.

- Antioxidant enzyme dependencies are:
    - glutathione peroxidase on selenium
    - superoxide dismutase on zinc, manganese and copper
    - catalase on zinc.

- Biotin
  Biotin is the important co-enzyme with Acetyl CoA for gluconeogenesis and energy production.[71]

- Butyric Acid
  Butyric acid is an agent for repair of the intestinal mucosa, when gut permeability is increased. It is a short chain fatty acid and occurs naturally in the gastro-intestinal tract of humans. It is a by-product of the fermentation of fibre by lactic bacteria, such as Lactobacillus acidophilus. Butyric acid should not be confused with gamma amino butyric acid (GABA), which is an amino acid.

  Note: glutamine is often prescribed for repair of the intestinal mucosa. However, in therapeutic doses it is often contra-indicated in Parkinson's Disease as it may manifest as an excitotoxin.

- Calcium
  Best absorbed in the organic form, It forms part of bone matrix. For bone integrity, vitamin D3 and magnesium should be considered together with calcium. There may be contraindications to calcium administration. Administration of therapeutic levels of Calcium may be contraindicated in Parkinson's Disease as excess entry into cells of Calcium and Sodium may de-stabilise neurones.

- Chromium
  This is implicated in Glucose Tolerance Factor (GTF).

- Co-enzyme Q10
  This nutrient plays a role in cell energy production and has possible implications in Parkinson's Disease.[72] It is also important in cardiovascular health.[73]

- Digestive enzymes
  Hydrochloric acid and pancreatic enzymes can be prescribed if deficiencies are medically demonstrated.

- Electrolytes
  Electrolyte combinations include sodium, potassium, calcium, magnesium and chloride amongst others

- Essential fatty acids (Omega 6 and Omega 3)[74]
  These enhance cellular membrane integrity, upregulate the immune system and are involved in the metabolization of anti-inflammatory and inflammatory prostaglandins series 1, 2 and 3.

- Folate
  This is involved in methylation and dopamine/adrenal metabolism.

- Gingko biloba
  This is an anti-oxidant which dilates capillaries and improves circulation.

- Iron
  Iron supplementation should be strictly medically monitored. Iron supplements should be taken at least two hours away from L-dopa administration and two hours away from taking zinc. There may be contraindications to iron administration generally and in Parkinson's Disease. Iron is best absorbed together with vitamin C. As the co-enzyme for tyrosine hydroxylase, it is essential for the metabolism of levodopa.

- Magnesium
  Deficiency can exacerbate muscle spasm and it is a co-enzyme in fatty acid and energy metabolism. Magnesium buffers excess intracellular Calcium

- Manganese
  This is a co-enzyme in the citric acid cycle and the antioxidant enzyme superoxide dismutase.

- Phosphatidyl serine[75] and Phosphatidyl choline[76]
  (See Note 2 on page 32) These are for cell membrane stability.

- Folate (folinic acid)[77]
  This is involved in the control of homocysteine and catecholamine synthesis. It has been linked with tetrahydro-biopterin[78] levels.

- Probiotics
  These include Lactobacillus GG/longum/plantarum/rhamnosus/salivarius/ acidophilus, bulgaricus, Bifidobacterium bifidum and other cultures of intestinal bacteria[79]. Each may have its individual indication and contraindication.

  Gut immunity may be enhanced by these friendly cultures of intestinal bacteria.

  Probiotics and saccharomyces boulardi supplemented alongside antibiotic therapy may aid the restoration of gut immunity.
  Brain – derived neurotrophic factor (BDNF) may be enhanced by probiotics, thus presenting possible positive influence on dopamine metabolism.[80]

- Prebiotics (nutrition for probiotics)
  Fructo-oligosaccharides and galacto-oligosaccharides may sometimes be indicated.

- Vitamin A
  This is implicated, amongst other functions, in the integrity of the gut mucosa and is an important antioxidant. Vitamin A and carotenes should only be administered in small doses if medically indicated. There may be cancer-related contraindications for taking these nutrients.

- **Vitamin B Complex**
  Vitamin B5 is involved in energy as well as adrenal metabolism. B vitamins are best taken as a complex. They play a role in mitochondrial energy production. Most L-dopa drugs contain a decarboxylase inhibitor, which prevents the further metabolism of L-dopa before entering the brain. As such, vitamin B6 can be used when taking "Madopar"[81]/"Sinemet"[82] and "Stalevo". It is an essential co-enzyme in the metabolic step between L-dopa and dopamine[83,84].

- **NADH (Vitamin B3 derivative)**
  This may stimulate L-dopa biosynthesis,[85] is the co-factor for the enzyme which forms tetrahydrobiopterin and is involved in cell energy metabolism[86].

- **Vitamin B12**
  This is implicated in neuromuscular function, methylation (methylcobalamine) and appears to be required for tetrahydrobiopterin synthesis.

- **Vitamin C**
  This is a coenzyme with dopamine ß hydroxylase and influences platelet aggregation.

- **Vitamin D3**
  Deficiencies have been detected in Parkinson's Disease[87]. It is implicated in bone metabolism.

- **Vitamin E**
  This is an antioxidant.

- **Zinc**
  This is implicated in many enzyme functions and protein metabolism. (See Note 1 below). It should be administered at least 2 hours away from iron.

NOTE 1: It is interesting to note the significant zinc deficiency in patients whether they take L-dopa or not. Significant zinc deficiency in the cerebrospinal fluid of Parkinson's Disease patients has been demonstrated in a controlled trial[88]. Zinc was also deficient in another controlled study assessing the nutritional status of patients with Parkinson's Disease[89].

NOTE 2: Curcumin[90]
Polyphenol active component of turmeric (Curcuma longa) with antioxidant, anti-inflammatory and anti-cancer properties, crosses the blood-brain barrier and may be neuroprotective.

References

1. Zigmund MJ, Bloom FE, Landis, SC, Roberts JL, Squire LR. Fundamental Neuroscience. San Diego: Academic Press; 1999. p.198

2. Thomas JH, Gillham B. Wills' Biochemical Basis of Medicine. Oxford: Butterworth Heinemann; 1989. p.159

3. Davidson VL, Sittman DB. Biochemistry. Baltimore: Lippincott Williams & Wilkins; 1999. p.393

4. Ramsey AJ, Daubner SC, Ehrlich JI, Fitzpatrick PF. Identification of Iron Ligands in Tyrosine Hydroxylase by Mutagenesis of Conserved Histidinyl Residues. Protein Science 1995;4:2082-6

5. Granner DK, Murray RK, Mayes PA, Rodwell VW. Harpers Illustrated Biochemistry. New York: Lange Medical Books/Mcgraw Hill; 2000. p.495-6

6. Granner DK, Murray RK, Mayes PA, Rodwell VW. Harpers Illustrated Biochemistry. New York: Lange Medical Books/McGraw Hill; 2000. p.267

7. Granner DK, Murray RK, Mayes PA, Rodwell VW. Harpers Illustrated Biochemistry. New York: Lange Medical Books/McGraw Hill; 2000. p.446

8. Mayeux R, Stern Y, Cote L, Williams JB. Altered Serotonin Metabolism in Depressed Patients with Parkinson's Disease. Neurology 1984;34.5:642-6

9. Pizzorno JE, Murray MT. Textbook of Natural Medicine Vol. 2 3rd Ed. St. Louis: Churchill Livingstone/Elsevier; 2006. p.1438-9

10. Datsis SA. Effect of Pyridoxine Deficiency on the Adrenal Cortex. Experimental Pathology 1991;43.3-4:247-50

11. Fidanza A, Bruno C, De Cicco A, Floridi S, Martinoli L. Influenza di Alte Dosi di Pantotenato di Sodio Sulla Produzione dei Corticosteroidi. Boll Soc Ital Biol Sper. 1978;54.22:2248-50

12. Patak P, Willenberg HS, Bornstein SR. Vitamin C is an Important Cofactor for Both Adrenal Cortex and Adrenal Medulla. Endocrine Research 2004;30.4 871-5

13. Pizzorno JE, Murray MT. Textbook Of Natural Medicine 3rd Edition. St. Louis: Churchill Livingstone/Elsevier; 2006. p.701-2

14. Schapira A. Mitochondria in the Aetiology and Pathogenesis of Parkinson's Disease. The Lancet Neurology 2008;7.1:97-109

15. Werner P, Mytilineou C, Cohen G, Yahr MD. Impaired Oxidation of Pyruvate in Human Embryonic Fibroblasts After Exposure to L-dopa. European Journal of Pharmacology 1994;263:157-62

16. Przedborski S, Jackson-Lewis V. Muthane U.Chronic Levodopa Administration Alters Cerebral Mitochondrial Respiratory Chain Activity. Ann Neurol 1993;34:715-23

17. Lehninger AL, Nelson DL, Cox MM. Principles of Biochemistry. New York: Worth Publishers; 1993. p.447-53

18. Voet D, Voet J. Biochemistry 3rd Ed. Hoboken: John Wiley & Sons; 2004. p.930-1

19. Voet D, Voet J. Biochemistry 3rd Ed. Hoboken: John Wiley & Sons; 2004. p.915

20. Bralley JA, Lord RS. Laboratory Evaluations for Integrative and Functional Medicine 2nd Edition. Norcross: Metametrix Institute; 2008. p.325

21. Leader G, Leader L. Parkinson's Disease The Way Forward 2nd Ed. London: Denor Press; 2003. p.78

22. Davidson VL, Sittman DB. Biochemistry. Baltimore: Lippincott Williams & Wilkins; 1999. p.393-4

23. Arthur JR, Beckett GJ. Thyroid Function. Br Med Bull 1999;55.3:658-68

24. Bland JS, Costarella L, Levin B, et al. Clinical Nutrition: A Functional Approach. Gig Harbor: Institute for Functional Medicine, Inc; 1999. p.168

25. Davidson VL, Sittman DB. Biochemistry. Baltimore: Lippincott Williams & Wilkins; 1999. p.394

26. Ponk K. Oxidative Stress in Neurodegenerative Diseases: Therapeutic Implications for Superoxide Dismutase Mimetics. Expert Opinion on Bilogical Therapy 2003;3:127-39

27. Werner P, Mytilineou C, Cohen G, Yahr MD. Impaired Oxidation of Pyruvate in Human Embryonic Fibroblasts After Exposure to L-dopa. European Journal of Pharmacology 1994;263:157-62

28. Pizzorno JE, Murray MT. Textbook Of Natural Medicine 3rd Edition. St. Louis: Churchill Livingstone/Elsevier; 2006. p.1088-9

29. Williams A, Steventon G, Sturmann S, Waring R. Heredity Variation Involved with Detoxification and Neurodegenerative Disease. Journal of Inherited Metabolic Disorders 1991;14:431-5

30. Jones DS, Quinn S. Textbook of Functional Medicine. Gig Harbor: The Institute for Functional Medicine; 2005. p.278

31. Jones DS, Quinn S. Textbook Of Functional Medicine. Gig Harbor: The Institute for Functional Medicine; 2005. p.361

32. Muller T, et al. Nigral Endothelial Dysfunction, Homocysteine and Parkinson's Disease. Lancet 1999;354:126

33. Hunot S, Hirsch EC. Neuroinflammatory Processes in Parkinson's Disease. Annals of Neurology 2003;53S3:S49 - S60

34. Erasmus U. Fats That Heal, Fats That Kill. Burnaby: Alive Books; 1993. p.276-8

35. de Lorgeril M, Salen P, Defaye P. Importance of Nutrition in Chronic Heart Failure Patients. European Heart Journal 2005;26.21:2215-7

36. Granner DK, Murray RK, Mayes PA, Rodwell VW. Harpers Illustrated Biochemistry. New York: Lange Medical Books/Mcgraw Hill; 2000. p.191

37. Erasmus U. Fats That Heal, Fats That Kill. Burnaby: Alive Books; 1993. p.276

38. Gardner M. Gastrointestinal Absorption of Intact Proteins. Ann Rev Nutrition 1988;8:329-50

39. Wang Q, Yu LG, Campbell BJ, Milton JD, Rhodes JM: 1998: Identification of intact peanut lectin in peripheral venous blood: Lancet 352:(9143): pps. 1831-2

40. Pusztai A: 1989: Transport of proteins through the membranes of the adult gastrointestinal tract - a potential for drug delivery?: Adv Drug Deliv Rev 3: pps. 215-28

41. Freed DLJ: 1991: Lectins in food- their importance in health and disease: J Nutr Med 2: pps. 45-64

42. Hadjivassiliou M, Gibson A, Davies-Jones GAB, Lobo AJ, Stephenson TJ, Milford-Ward A: 1996: Does cryptic gluten sensitivity play a part in neurological illness?:Lancet 347: pps. 369-71

43. Furnass JB. Types of Neurones in the Enteric Nervous System. Journal of the Autonomic Nervous System 2000;81:87-96

44. Riley PA. Free Radicals in Biology – Oxidative Stress, and the Affects of Ionizing Radiation. International Journal of Radiation Biology 1994;65.1:27

45. EPA Health Advisory for Simazine (Herbicide): EPA HA d250: Herbicide – nervous system effects including tremor. Environmental Protection Agency, Office of Water 2004:225,277

46. Smeh NJ. Herbicide (1-methyl-4-phenyl-1,2,3.6-tetrahyropyrinine) is implicated in both stiffness and weakness in Parkinson's disease. Save the Children and Yourself – A Guide to a Future and Healthier Generation by Avoiding Toxins in Today's Food and Water. Alliance Publishing Company 1996:189,226

47. Ngim CH, Pevathazan G. Epidemiological Study on Association Between Body Burden Mercury Level and Idiopathic Parkinson's Disease. Neuroepidemiology 1989:8:121-41

48. Blaylock R. Excitotoxins – The Taste that Kills. Santa Fe: Health Press; 1997. p.39-43

49. Choi DW. Glutamate Neurotoxicity/A Three Stage Process/Neurotoxicity of Excitatory Amino Acids. FIDA Research Foundation, Symposium Series. Raven Press 1990:4

50. Kempster PA, Wahlqvist MC. Dietary Factors in the Management of Parkinson's Disease. Nutrition Reviews 1994;52.2

51. Leader G, Leader L. Parkinson's Disease Reducing Symptoms With Nutrition And Drugs Revised Ed: London: Denor Press; 2009. p 66-67

52. ABPI Compendium of Data Sheets and Summaries of Product Characteristics. Datapharm Publications LTD 1999-2000;1169

53. ABPI Compendium of Data Sheets and Summaries of Product Characteristics (2000): Datapharm Publications Limited, London, UK: p. 1183

54. Electronic Medicines Compendium (eMC): 2008

55. Dr Geoffrey Leader MB ChB FRCA and Lucille Leader Dip ION BANT: 2009: Parkinson's Disease - Reducing Symptoms with Nutrition and Drugs: Denor Press, London, UK: p.10

56. J Hywel Thomas PhD FIBiol and B Gillham PhD: 1989:Wills' Biochemical Basis of Medicine: Butterworth Heinemann: Oxford, UK: p.417

57. Campbell RR, Hasinoff B, Chernenko G, Barrowman J, Campbell NR. The effect of ferrous sulfate and pH on L-dopa absorption: Can J Physiol Pharmacol: May 1990;68(5):603-7.

58. ABPI Compendium of Data Sheets and Summaries of Product Characteristics (1999-2000): Datapharm Publications Limited, London, UK: p. 1105

59. ABPI Compendium of Data Sheets and Summaries of Product Characteristics (1999-2000): Datapharm Publications Limited, London, UK: p. 1104

60. Malcolm SL, Allen JG et al: 1987: Single Dose Pharmaconetics of Madopar HBS in Patients and Effect of Food and Antacid Absorption of Madopar HBS in Volunteers: European Neurology: Vol 27. Suppl 1. 1987

61. www.medicines.org.uk

62. ABPI Compendium of Data Sheets and Summaries of Product Characteristics (1999-2000): Datapharm Publications Limited, London, UK: p. 1610

63. Dr David Perlmutter MD: 2000: BrainRecovery.Com: Perlmutter Health Center, Naples, USA: p26

64. Oja SS, Jankay R, Varga V, Saransaari P: 2000: Modulation of Glutamate Receptor Functions by Glutathione: Neurochem Int Aug-Sept 37 (2-3): pps. 299-306

65. Steffen V et al: 1995: Effect of Intraventricular injection of l-methyl-4-phenylpyridinium protection by acetyl-L-carnitine: Human Exp Toxicol: (14): pps. 865-871

66. European Journal of Pharmacology 263 (1-2): 1994: September 22: Impaired oxidation of pyruvate in human embryonic fibroblasts after exposure to Ldopa: pps.157-62

67. C W Olanow: 1989: Attempts to obtain neuroprotection in Parkinson's Disease: Neurology 49: Supplement 1: S26-S33

68. Pong K: 2003: Oxidative stress in neurodegenerative diseases, therapeutic implications for superoxide dismutase mimetics: Expert Opinion on Biological Therapy: 3: pps. 127-39

69. Juurlin H, Paterson PG: 1998: Review of oxidative stress in brain and spinal cord injury, Suggestions for pharmacological and nutritional management strategies: B: J Spinal Cord Med Oct: 21 (4): pps. 309-334

70. Dr David Perlmutter MD: 2000: BrainRecovery.Com: Perlmutter Health Center, Naples,USA: p25

71. Victor Davidson, Donald Sittman: 1999: Biochemistry: Lippencott, Williams, Wilkins USA: p325

72. Shults C et al: 1999: A Possible Role of Coenzyme Q10 in the Etiology and Treatment of Parkinson's Disease: BioFactors (2-4): pps. 267 - 272

73. Langsjoen H, Langsjoen P et al: 1994: Usefulness of Coenzyme Q10 in Clinical Cardiology: A Long Term Study: Mol. Aspects Med: 15 Suppl. S165-S175

74. Yehuda S,Rabinovitz S et al: 1998: Review: Fatty Acids and Brain Peptides: Peptides: Vol 19. No 2: pps. 407-419

75. Dr David Perlmutter MD: 2000: BrainRecovery.Com: Perlmutter Health Center, Naples, USA: pps 23-24.

76. Cui Z, Houweling M, Review: 2002: Phosphatidyl choline and Cell Death: Biochemica et Biophysica Acta: 1585:pps.87-96

77. Fernstrom JD: 2000: Am J Clin Nutr: 71(suppl): pps. 1669 S-73S

78. Hamon CG et al: 1986: The Effect of Tetrahydrofolate on Tetrahydrobiopterin Metabolism: J Ment Defic Res 30: pps. 179-183

79. Leon Chaitow ND DO and Natasha Trenev: 1990: Probiotics: Thorsons: An Imprint of HarperCollins Publishers, London, UK : pps. 24-25

80. Berton et al: 2006: Nature Reviews Neuroscience 7:pps.137-151

81. ABPI Compendium of Data Sheets and Summaries of Product Characteristics (2000): Datapharm Publications Limited, London, UK: p. 1183

82. Electronic Medicines Compendium (eMC): 2008

83. Dr Geoffrey Leader MB ChB FRCA and Lucille Leader Dip ION BANT: 2009: Parkinson's Disease - Reducing Symptoms with Nutrition and Drugs: Denor Press, London, UK: p.10

84. J Hywel Thomas PhD FIBiol and B Gillham PhD: 1989:Wills' Biochemical Basis of Medicine: Butterworth Heinemann: Oxford, UK: p.417

85. Volc D, Birkmayer JG, Vrecki C, Birkmayer W: 1993: NADH - A new therapeutic approach to Parkinson's Disease. Comparison of oral and parenteral application: Acta Neurologica Scandinavica Suppl. 146: pps.32-35

86. Nadlinger K, Westerthaler W, Storga-Tomic D, Birkmayer J.G.D: Extracellular metabolisation of NADH by blood cells correlates with intracellular ATP levels: November 2002: Biochimica et Biophysica Acta (BBA)/General Subjects: Volume 1573, Number 2, 14: pps. 177-182

87. Y Sato, M Kikyuama, K Oizumi: 1997: High Prevalance of Vitamin D deficiency and reduced bone mass in Parkinson's Disease: Neurology 49(5): pps. 1273-79

88. Ward NI et al: 1988: Trace Element Status of Cerebrospinal Fluid of Individuals with Neurological Diseases by ICP-MS Trace Elements Analysis in Diagnosis and Pathological States: Vol 5: Proceedings of the 5th International Workshop in Nuremburg, Rep of Germany: Trace Element Analytical Chemistry in Medicine and Biology: Walter de Gruyter & Co, Germany / USA: pps. 513-550

89. Abbot RA et al: 1992: A Diet, Body Size and Micronutrient Status in Parkinson's Disease: European Journal of Clinical Nutrition 46(12): pps. 879 - 884

90. Mythri RB1, Bharath MM:Curcumin:a potential Neuroprotective Agent in Parkinson's disease: 2012: Curr Pharm Des18(1):91-9.

Bibliography

Electronic Medicines Compendium (eMC) at http://emc.medicines.org.uk: Datapharm Communications Ltd; 2009.

Leader G, Leader L. Parkinson's Disease Reducing Symptoms With Nutrition And Drugs Revised Ed: London: Denor Press; 2009.

Bralley JA, Lord RS. Laboratory Evaluations for Integrative and Functional Medicine 2nd Edition. Norcross: Metametrix Institute; 2008.

Jones DS, Quinn S. Textbook Of Functional Medicine. Gig Harbor: The Institute for Functional Medicine; 2005.

Lieberman M, Marks A.D, Smith C. Marks' Essentials Of Medical Biochemistry, A Clinical Approach: Baltimore: Lippincott Williams & Wilkins; 2007.

Bender D.A. Nutrition And Metabolism 4th Ed: Boca Raton: CRC Press; 2008.

Granner DK, Murray RK Mayes PA, Rodwell VW. Harpers Illustrated

Biochemistry: New York: Lange Medical Books/Mcgraw Hill; 2000.

Pizzorno JE, Murray MT. Textbook of Natural Medicine Vol. 23rd Ed. St. Louis: Churchill Livingstone/Elsevier; 2006.

Voet D, Voet J. Biochemistry 3rd Ed. Hoboken: John Wiley & Sons; 2004.

Leader G, Leader L. Parkinson's Disease The Way Forward 2nd Ed. London: Denor Press; 2003.

ABPI Compendium of Data Sheets and Summaries of Product Characteristics. Datapharm Publications Ltd 1999-2000.

Davidson VL, Sittman DB. Biochemistry. Baltimore: Lippincott Williams & Wilkins; 1999.

Perlmutter D. BrainRecovery.Com: Naples, Florida: Perlmutter Health Center; 2000.

Davidson VL, Sittman DB. Biochemistry. Baltimore: Lippincott Williams & Wilkins; 1999.

Blaylock R. Excitotoxins – The Taste that Kills. Santa Fe: Health Press; 1997.

Erasmus U. Fats That Heal, Fats That Kill. Burnaby: Alive Books; 1993.

Thomas JH, Gillham B. Wills' Biochemical Basis of Medicine. Oxford: Butterworth Heinemann; 1989.

Geoffrey S Bland PhD et al: Clinical Nutrition - A Functional Approach:The Institute for Functional Medicine Inc, Gig Harbor, WA, USA:1999.

Victor L Davidson, PhD, Donald B Sittman, PhD: Biochemistry: The National Medical Series for Independent Study, Harwal Publishing, Baltimore, USA, 1994

J Hywel Thomas, PhD, FIBiol & Brian Gillham PhD: Wills' Biochemical Basis of Medicine, Second Edition: Butterworth Heinemann Limited, Oxford, UK, 1993

Stephen A Levine PhD & Parris M Kidd PhD: Antioxidant Adaptation - Its Role in Free Radical Pathology: Allergy Research Group, California, USA, 1994:

Udo Erasmus: Fats that Heal, Fats that Kill: Alive Books, Burnaby BC, Canada, 1993

Nigel Plumber BSc PhD: The Lactic Acid Bacteria - Their Role in Human Health: Biomed Publications Limited, Shirley, UK, 1992

Michael Murray ND and Joseph Pizzorno ND: Encyclopaedia of Natural Medicine: Prima Publishing, Rocklin, CA, USA, 1991

Michael Ash, BSc DO ND Dip ION MBANT: Atypical Depression, The Immune System, Probiotics and Clinical Application: The Stressed Gut -The Stressed Brain: Royal Society of Medicine Presentation: Food and Health Forum: London, UK, 2008

Mitchell Bebel Stargrove, Jonathan Treasure, Dwight L McKee: Herb, Nutrient and Drug Interactions: Moseby/Elsevier: 2008

Liener IE et al: The Lectins - Properties, Functions and Applications in Biology and Medicine: Nutritional Significance of Lectins in the Diet: Academic Press, Orlando, USA, 1986

Index

41

www.ingramcontent.com/pod-product-compliance
Lightning Source LLC
Chambersburg PA
CBHW062030210326
41519CB00060B/7378